Table

Chapter 1: Purpose	2
Chapter 2: Journey	4
Chapter 3: Miriam's Letter	8
Chapter 4: Reaching Capacity	10
Chapter 5: Dangers, Toils, and Snares	13
Chapter 6: Meet the Original Family	33
Afterwards: Going Home and Continuing On	40
All About: Author and Organization	41

__1__
Purpose

learn to do good;
seek justice,
   correct oppression;
bring justice to the fatherless,
   plead the widow's cause.
Isaiah 1:17

A group of townspeople were gathered around the little lifeless lump on the front veranda of a tiny grocery store in the dusty town of Maralal, Kenya. I was walking up to the scene, dreading having to deal with the death of a little street boy. He looked lifeless. I thought he had possibly only just died, because he was still so limp. They would gently lift the small boy by the arms and shake him, then drop him back to the cement pad. No response.

Just the night before, I was in my motel room praying that God would show us a way to help some of these street boys in Maralal, Kenya. There were boys just wandering around, collecting used bottled water containers to sell for a shilling, digging through scraps to find food before the dogs or cows did, or huddling up in a storefront to try and rest. Many of the boys were under the age 10.

The very next morning I found myself at the entry to the grocery store with a boy of about 6 years, lying responseless in his own urine. I was relieved to discover he was just suffering from hypothermia. People in the town knew me by this point in our tenure among the Samburu and knew that I have a background in orphan care and child counseling and so they began demanding that I take action with him with such heart piercers as, "God has called you to help boys like him, so help him!" and "He is your expertise, only you can help him." I didn't know what else to do but see him as an answer to my prayer to God. After all, why did we move to the bush of Kenya in the first place? It was to establish a family, built for the neediest of the needy children, just like the boy at my feet.

2

Journey

But we are not of those who shrink back and are destroyed, but of those who have faith and preserve their souls. - Hebrews 10:39

Satan hates it when we steal children that he thinks are rightfully his. The street boys, the forgotten, mistreated orphans are all within his domain, or so he thinks. We were heading into a remote land where alcohol had claimed most living fathers who were still around the kids, other dads ran away to drink in peace, a good number, though, were claimed by AIDS or violence. The future of a boy in Samburu was to be a drunk. The future of a girl was to let a drunk man take advantage of you and run away. So, Satan had the offspring of these dead and dying, or deadbeat parents in his pocket. No Christian children's home existed. No Christian churches really cared. But then things changed. Child Help International was sending us into Satan's pocket to reclaim what is rightfully God's. However, Satan wasn't letting us in without a fight.

Our assignment in Kitale had finally come to a finish. It had so many challenges and blessings. My wife, Stacy, had grown close to the owner of the lodge where our small, but beautiful house was located. I had been overseeing a frustrating construction project and equally frustrating leadership training program. Those would only serve as a preparation for the challenges to come. Our oldest boy, Daniel, was 12 and always vacillated between loving Kenya and hating it. Miriam, our 9 year-old daughter, absolutely loved life in Kenya. Micah, our little 7 year-old, was pretty much fine with everything and enjoyed the adventures. None of us would've thought we were prepared for what was to come. And getting out of Kitale was just the start.

We'd been attempting to leave Kitale since the beginning of that week, but everyday there was a delay. My truck (a.k.a. "DOA") overheated, blew a gasket, burnt the spark plugs, and shorted out the starter. That was Saturday, Sunday, Tuesday, Wednesday, and Thursday. During that time, we had to squeeze in the court hearings and legal errands concerning our foster daughter, Blessed, a small girl Miriam had saved.

Suddenly, there was a shortage of funds because of the vehicle repairs and court costs. It appeared all hope to move to Samburu district would be lost. In my view, it was quite the opposite. We had a mission that God has called us to do. No one or thing could stop what God had willed. So, when a trip is delayed, why despair? We baptized 16 kids from the Haven Children's Home in Kitale, we made the children's home there a little better, and we established new relationships for His glory. The enemy hadn't really slowed us down, he'd allowed us to have even more victories in the very place we were intending to leave.

When the truck started working and then refused to ever start again, we gathered in prayer against the enemy and proclaimed the victory we would soon experience. It was later discovered that the mechanics caused most of the problems and were the ones fixing one thing and breaking another in hopes of gleaning as much money from the Americans as they could.

But then the victories began coming. I decided the truck would never get fixed with the crooks in town, so we unpacked it and left it for Haven Children's Home to sell and split the money with us or eventually fix on their own and keep for their needs. We would be relying on God to transport us around the bush of Kenya and not the truck I had bought for that purpose. At the same time, the Kitale district judge ruled in our favor and placed Blessed in our legal custody until the age of 18.

We hired a van to get us all to Samburu County and arrived safe, happy, and a lot poorer after all the repairs and travel expenses. We would stay in town for five days and then we move to our tented camp for the next 5 months to set up the children's home in partnership with Loosuk Christian Church and Domain Christ Ministries.

We trusted God to refill the financial cup, but it is always a little unnerving to arrive in the middle of nowhere Kenya with a hundred bucks to get you along. We had 5 days to pay our lodging expenses and purchase materials to put up our new tent homes in a safe and effective manner. We sent the word out to pray along with us for the funds to get our ministry started. Nevertheless, having finally arrived safely, with Blessed in our Custody were victories enough for us. Having a decent place to stay while I constructed our home was icing on the cake.

Daily, I whisked back and forth with a motorbike taxi to our new tented home site about 1 mile from Loosuk village. Loosuk was a tiny village not on any map. A dusty old road separated a line of storefronts, looking not so different from the main street you'd see in an old western film. To reach our homesite required a 45 minute ride on a terribly rocky, rough, washed out dirt road that climbs steeply to mountain tops and dips severely back into stony valleys.

Each day we worked to prepare the site that would first be our home and then a home to many kids who had lost everything. With my helpers, we set up UN refugee tents, an outdoor american-style toilet, and a structure to put the tents under.

Meanwhile, God was working on the financials and some Christians were willing to allow us to borrow some funds to get us going. Within a week that was repaid and we were finally going again on our own, thanks to the Lord's provision.

A couple days delayed due to the funding problem, but with a bit more money to help us along, we loaded a pick up truck with all our belongings and bounced our way to our tented homestead. Here we would stay, building a children's home, a church body, and a new Christian family for the next 5 months.

3
Miriam's Letter

"I have been a sojourner in a foreign land." -Exodus 2:22

What better way to get a glimpse into the life in the bush than through the eyes of child? The following is a letter our daughter wrote to her cousin back in Idaho. Miriam was 9 years-old at the time.

Hi how are u? I am fine. We live in Africa in Loosuk, north of Maralal. Our house is 2 big tents in the middle is the kitchen and dining room. Each tent has 2 rooms. I am on the right and the boys on the left. Mom and Dad are on the left of their tent and the right is the tv room. We cannot watch tv yet, because we don't have electricity so every night my dad reads from a book. Only one chapter. The book is Princess Bride.

We cook on a thing called a jiko, it is a small clay and metal bowl that holds charcoal in it. It has some holes where ashes drop through the bottom. It takes a long time to boil water, but in the mornings Phyllis, our maid, makes coffee, chai, and sets out oatmeal, cereals, and makes toast. My dad and mom boil milk for the oatmeal. The milk comes straight from cows that walk around outside our tent. We have to boil it so we can drink it and not get sick. The water here has to be put through a filter or boiled so you can drink it and not get sick. We get our water from Loosuk a town near us.

Someone walks or takes their motorcycle to fill our cans with water.

Us kids sleep in sleeping bags on mattresses. For lunch we have Kenyan food like Ugali, which is like play-dough with no taste, rice, cabbage, and potatoes.

We live on a savannah and if our guard, Zakayo, leaves the gate open zebras come in and eat the grass. They look like painted donkeys, that's why I think in the language here their name means "striped donkeys".

Every day after we read the Bible, my dad goes and meets with government officials, church people, or builds on our house. Sometimes he gets food for us if we really need it. Today he met with the Chief of Loosuk and helped put iron sheets up on our roof. Us kids were jumping on the tarps and doing schoolwork. Soon after I finished, and Daniel was still stuck in math, we went out for a walk and explored different paths and named them.

I think being in Kenya is fun. I hope to see you soon.

Your best friend,cousin,

Miriam

P.S. There is no cheese here!

4

Reaching Capacity

Lord, I know that a person doesn't control their own life. They don't direct their own steps. -Jeremiah 10:23 (NIrV)

We never planned on being foster parents again. We had 19 youngsters come in and out of our home in Nampa, Idaho back in 2003 through 2004. It was a wonderful experience that would last forever in our minds. However, taking it on again was not really on our minds at all. It really started the December before arriving in Samburu.

The director of our Orphan Care Ministry in Kitale, Kenya told us of a poor woman, struggling to care for orphaned babies and toddlers. He said with sadness that a baby had starved to death there just the week before. Stacy decided we should visit so she could provide some medical relief for the babies and toddlers. I simply wanted to talk with the young lady running the orphanage and explain that there is a rescue center in town for at-risk orphans that would be better than letting them starve in her home.

When we arrived, the director was quick to dismiss my plea for shutting the place down. Stacy didn't notice because she was checking children one by one. Then Miriam brought Stacy to a tiny figure of a girl who was fading fast. Too weak to lift her arms, barely able to turn her head, and hardly able to speak. Her eyes were expressionless, but still seemed to cry out for rescue. Stacy couldn't say no. Nor could I. Little Blessed Benta Akiru joined our family as our first foster child.

Like all our foster children, even in the States, she brought many challenges. More than we had faced in some time. She had been raped, starved, and neglected. Now we had to love her to healing emotionally, spiritually, and physically. God gave her to us first, because she needed the full nine months to heal.

Soon into our time in Samburu, we were fostering 3 street boys, and 2 orphaned siblings. Robert came first, whom you read about at the beginning of this story. He was a feisty, self-reliant little street boy of seven. He refused to use the toilets, ran away, fought with every adult, and cried until he had no more tears. After 2 weeks of of these struggles, finally something in him broke and we had him living healthy and happily with us.

We soon, and just as unexpectedly, added another boy when I was back in Maralal to make arrangements for a committee leadership meeting. Two committee members were also in town gathering their own supplies for the meeting. In tow? A street boy, naturally! They said that he was from their village and even attended their church but that his father died of AIDS and since the father's death his mother ran off to serve as a prostitute to support her alcohol addiction. The little guy began starving and decided the small city was his best chance at survival. They said they found him begging for food and took him to see if I could rescue him.

I looked over the boy hiding under layers of dust in clothes many sizes too large. His big eyes filled with a mix of curiosity, hope, and a touch of fear. What could I say? I had David scoop the boy up onto his motorbike and bring him safely into our fold.

Philip brought a new set of violent challenges for our rehabilitation process. Fits of rage would turn to a boy curled up on a lap, softly whimpering and desperately eking out every second of the human contact.

Robert and Philip, along with ten year-old Manuel and his four year-old sister, Sheila, filled our refugee tents unexpectedly. Yet there was room for one more street boy in our make-shift children's home. Maybe God would lead him to us like these other youngsters.

Our home was filled with laughter, joy, fights, stealing, and emotional breakdowns. I counseled, we prayed, and lovingly disciplined them toward a walk with Christ. Prayer warriors across the States propped us and our newly hired staff up in prayer as we worked with these young, broken souls. From Satan's coffers, four had been repossessed.

5

Dangers, Toils, and Snares

...and call upon me in the day of trouble; I will deliver you, and you shall glorify me. Psalm 50:15 (ESV)

Life seemed to be smooth and heading toward a beautiful finish for our assignment among the Samburu. Things got really exciting when we received a special donation for a specific purpose, to buy a motorbike to zip us around the outback! It arrived on a Wednesday. I began lessons on Thursday. It would soon become as great a companion as my trusty dog, Matthew.

Aside from that excitement, the children we were fostering before the home opened were adjusting very well and starting to enjoy having a family and structured life of rules, consequences, meals, and playtime.

Things turned tough one night when I finally mustered the courage to announce that Stacy and I had to go to Nairobi to renew our visas for one last time. I held two sobbing boys and Stacy comforted a sad Miriam and Micah. Our own children didn't want to be left in the bush and the two boys didn't want to lose another "father" in their life. Despite the heartache, we departed early one morning before the mists had cleared the savanna. After 5 hours on a motorbike, a quick hour car jaunt, and 3 more by matatu (Kenya's van transport) we reached Nairobi safe and sound.

The next day we received our visas from a kind official at Immigration who allowed us to renew without going through all the hoops and red tape. It was just a small "service fee." We were thankful to God, since we have yet to receive missionary visas and we were unsure how they would react to renewing visas for the 3rd time. We went from palpitating hearts to prayers of praise not to be forced to endure days of bureaucracy, or worse denial and deportation. God had our backs and we were in for smooth sailing. Or so we thought. In reality, 2 Corinthians 4:8-10 would soon be illustrated all too clearly.

"We are afflicted in every way, but not crushed..."
                                         2 Corinthians 4:8a

They counted four seizure like experiences and said I then went insane. I slipped in and out of consciousness, started speaking only in Spanish, and would lapse into another fit of chills. Cerebral malaria made that particular Sunday a truly terrible day. After preaching in the service, I went to bed and progressively got worse. Stacy found me convulsing in the bed and immediately realized I had malaria, the deadly cerebral version.

Philip entered my compartment in the refugee tent and collapsed in a sobbing heap on my chest. His sister, once his last care provider, died of cerebral malaria. Here was his new care provider at death's door.

Amazingly, through much prayer and Stacy's uncanny medical knowledge, I regained consciousness and progressed back to what is considered a normal mental state for me, though confusion and randomly speaking in Spanish was still occurring.

Stacy, though, had to leave us for a short while. Our foster girl from Kitale was due for another court hearing in Kitale in two days. Stacy, Blessed, and an older ministry leader, Dominic, had to leave first thing in the morning. While I lay suffering in my bed, I had no idea what suffering Stacy and Blessed would soon endure.

"persecuted, but not forsaken..."
2 Corinthians 4:9a

A terrible series of problems would lead Stacy to a scene no person ever wants to witness, let alone be in the midst.

Six months back, I had been preparing to leave Nairobi to return to Kitale on the 9:00 am van shuttle from just outside downtown Nairobi. However, a sickening pit in my stomach soon changed my mind. I called Stacy at our house in Kitale and informed her I was physically unable to make the trip by van shuttle that day. The next morning's newspaper told the story of a homemade bomb going off at that shuttle stage at the time I might have been there. Al-shabaab took the glory and threatened even more violence. Our white skin and missionary endeavors made us even more of targets across the country, not just for Al-shabaab attacks, but more likely robberies and mob violence. It seemed unlikely, but I still made some hard rules about public transport in Kenya. One, never pick up the shuttle or matatu from the stage. Two, never get off at the destination's stage. Three, never travel after dark. All of those rules were ignored by Stacy's travel companion. Stacy would pay the price.

Early in the morning, Stacy, Blessed, and Dominic began a 16 hour journey that would require 4 crowded matatus over terrible roads. Stacy endured the hard, worn seats, the hot, stuffy atmosphere, the peering crowds, and even being vomited on by a local who took the journey worse than my wife. At last they arrived in Kitale. But their arrival was in at the matatu stage stop in the dilapidated slum of Kipsongo, where Blessed was born.

Stacy was immediately noticed and there were more than just gawkers around this late at night. Stacy hurried to get herself and Blessed onto a motorbike taxi, while Dominic began discussing the plans for the night. Upon seeing Stacy with a little African girl on the same motorbike, a man began shouting that Stacy was kidnapping one of the children from Kipsongo. Shouting ensued. Gestures went from angry to violent. Nearly 20 men, filled with an alcohol-fueled rage rushed at Stacy and Blessed and began pulling Blessed's arms trying to rip her from Stacy. Stacy yelled at the men that she had custody and court papers. Blessed's tear filled screams of fear and pain were barely heard amidst the incessant slurred yells of the men pulling her away from Stacy. Somehow, in a moment of strength, courage, and sheer willpower, Stacy began hitting the hands that tore at Blessed. Stacy then pulled Blessed close and curled herself over top of Blessed in order to absorb any more blows and pulls.

Not soon enough, police arrived on the scene and began breaking up the angry mob. They plucked Stacy out of the mob and arrested her for inciting a riot. Thankfully, Blessed and Dominic were allowed to be a part of the proceedings at the city jail.

From inside the police station, Stacy dared to begin texting. First, she texted me on my own bed of suffering. The words "I've been arrested" somehow knocked even the slightest confusion from my malaria stricken head. Immediately, I called our attorney in Kitale, waking him up to confront the police. Second, Stacy texted our friend Theresa, the owner of Karibuni lodge where we had a house for 7 months. Theresa was friends with the chief of police. Inundated with threats from the Chief and a lawyer, the police had no choice but to relent and release Stacy. They escorted her, Dominic, and Blessed in their police truck safely within the guarded gates of Karibuni Lodge.

The court hearing lasted mere minutes and soon Stacy, Blessed, and Dominic were returning back to us in Loosuk.

"struck down, but not destroyed..."
2 Corinthians 4:9b

I don't usually wear a helmet because I drive so slow. It is hard to get up to a good speed when you are bumping up big rocks to get to the top of the next hill. Usually, I'd just cut the engine to get down the next dangerous decline. I wasn't against the helmet, it just seemed unnecessary and could block some vision of the rocks, holes, and dirt slides. Sure, I'd been in a few different "accidents" on the bike. Usually from slipping in the mud or down a dirt slide. Those times the worst would be a possible bruise if anything. Today was different. A cop had shown up in Maralal and was enforcing the helmet law. So, I brought it along and tried to get another passenger wear it, but he refused. I had 3 passengers on the motorbike to take across the mountains to town. But none of them would wear it. So, the helmet ended up where it belongs, on my head.

The drive was bumpier than normal. We took the less treacherous back dirt road built up by the British over half a century ago. From there we got to go through my favorite motorbike area in the world. It was just motorbike path crisscrossing motorbike path through the savanna, along the base of a rocky cliff. You would dip, turn, duck under a fallen tree, and arrive at a slim road that led right up into Maralal. It was quite a magical place.

I drove us to that big, rough hill that leads into Maralal and dropped off my passengers to walk since it would require me to do some fancy work to off road the bike up. I had misjudged a few things. My bike was now 350 pounds lighter and a lot more powerful. The road itself, though, was washed completely away and only a 6 inch ridge remained. Beside that skinny ridge a thorn and barbed wire fence. Down the 6 inch ridge was a 3-5 foot deep canyon where the road once was. I gunned it up the hill, maneuvering carefully, but surprised at the speed I quickly accumulated. The path narrowed, the middle canyon threatened. I didn't clear the ridge or jump the small canyon. Instead, my right handlebar made a loud snap as it clipped the barbed wire and I was thrown from the bike.

In an instant, the world's noises became a loud ring. Vibrations went through my entire body and seemed they would never end. Suddenly, I was aware of what had happened and that my daughter had witnessed it. Adrenaline rushed through my body. I jumped up as quick as I could. I had flown to near the top of the hill, the bike was stopped from landing on top of me by falling into the small canyon, sticking partway out of the top of it. I grabbed the bike and managed to rev it out of the pit and up to the main road going into Maralal. My companions rushed up to me, and I assured them I was okay. My daughter did not see what had happened, but heard the fearful talk from our friends and saw how they rushed to me as fast as they could.

As they climbed back onto the bike, I tried to rev the engine and get us moving again. My adrenaline was subsiding and the injuries were starting to make themselves known. I could no longer turn my wrist. I asked our friend to take over driving, and asked him to take me to the little muslim clinic in town.

We pulled up to the clinic and I told my companions to do what they needed to and to meet me at the restaurant next door. By this time my hands were covered in blood and I was starting to have trouble thinking straight.

With Miriam's help, I removed the helmet. It was a strong, expensive helmet. When I looked at it, though, it looked old and cheap. The top was scraped up and tiny dents were made by the rocks. According to the two men who witnessed it, I flew from the bike landing head first. My arms had flailed and smacked the ground. I had hit the top ridge of the canyon in the center of the road and slid up and out. My head and hands plowing the way. My leather jacket was covered in dust with tiny scrapes along it.

The medic cleaned my hands. The right hand stung and ached with the movement, but it was my left hand looked like ground meat.

Only the left mirror broke on the bike but I headbutted the ground and shredded my left hand, contused both hands, suffered a concussion, and have a hairline fracture or three in my right hand. There are no x-ray facilities for 5 hours, so everything was guessed at a small muslim hospital. Sunday was a day of pain, but today I am up and about again thanks to prayer and heavy meds! Best part? Hospital bill and meds = $8.33!

"perplexed, but not driven to despair..."
2 Corinthians 4:8b

I felt a little like Ronald Reagan during the air traffic control strike. Our entire construction crew went on strike demanding more money since I am supposedly a rich American. I fired them all and hired an entirely new workforce.

This all happened on a very perplexing week. The construction workers were instructed to get the work mostly done, if not finished during the week I was gone on a quick trip to Nairobi to pick up our foreigner Resident ID Cards. Instead, they discussed life, took time off, ate free food, then left on strike, holding out for more pay. While Stacy was left dealing with that in my absence, she was also dealing with a runaway young teen girl who was nearly killed by her uncle- who happened to be employed as our security guard (I fired him the day before firing our construction workers). While dealing with our residency issues, I went to visit a real doctor who said I had a "shattered wrist" or at least a "bad break" in my right hand. It was safe to say Satan was making our last month in Kenya as challenging as possible.

I returned home to a world of chaos. No construction workers, a teen girl in hiding, a guard I needed to fire, and a street boy who decided attempting suicide would be the best way to punish me for leaving him.

After some prayer, and feeling the power of our supporters prayers behind me, I set to work. First, sitting down with the leader of our partner ministry and our guard, releasing the guard from duty and encouraging him to go and grow spiritually. Second, calling up the construction workers and firing them. Third, hiring a completely new crew who set to work the next day. Fourth, uniting the teen girl with her care providers after a lengthy discussion on abuse and their liability. Fifth, sitting down with the former street boy and his soon to be Dad (the church pastor and orphan care ministry director) to discuss the realities of my leaving soon and his need to embrace the rest of the family and not just me. Needless to say, it was a difficult few days. But in it all, we persevered! Construction was back on track, the teen girl was moving toward healing and a healthier lifestyle, the former street boy, Philip, was indeed trying to embrace the rest of the family. All the perplexing troubles seemed to fade into the past. Until I went back to the ATM.

Our money had dried up and we needed to finish the roof! We were about a thousand dollars short of being completely finished with construction. Somehow, our well of funds had run dry. Now we were back to praying and waiting.

It wasn't even an entire 24 hours until I received an email from our forwarding agent that an unexpected amount came from two camp sessions at White Mills Christian Camp in Kentucky and she had deposited it immediately. The total came to just over a thousand dollars.

"so that the life of Jesus may also be manifested…"
2 Corinthians 4:10b

Finally, it was the big committee meeting to get the leaders squared away with the protocol and standards for their new orphan care ministry. It was months in the making, since translating the manual began in May and was completed...in late September. Nothing like cutting it close! The meeting was a long but beneficial and well received experience. It was great to see 5 churches so eager and unified for the fatherless in their community!

Of course, being on African time it ended late, which meant they couldn't do the 2 hour walk back to their homes in the bush. After a week of intense work on the house, translation, and being a foster dad to 3 clingy street boys, I was exhausted and my fibromyalgia was acting up big time. But I had to head out and figure a cheap way to get all the committee members home, buy some construction materials, and get it sent to the new home. I knew I was unable to make the trip physically and just wanted to lay down. After everything was dealt with and on their way, I drove my motorcycle back to the hotel. Only to hear little frantic voices yelling "pasta net!" (Pastor Nate). A horde of street boys came running toward me. We've lost a little 10 year-old who has been on the streets for a couple months. They don't know where he is and wanted me to fix the problem. Seeing their frantic state and needing to talk to them calmly I made a big deal about bugs in their hair and set off to get their heads shaved while they related to me the events since they'd last seen me.

After the haircuts, we wandered the alleyways until dark, when it was no longer safe for me to be out or be seen with them. Exhausted, I stopped to buy them a roll and milk carton before they headed to their storefronts to sleep. There at the storefronts was their friend, apparently also exhausted from a long day on the street.

Just ten days after the day of exhaustion and here we were on a cloudy, but joyous Sunday. Water in the local watering hole was deep enough for baptisms and the children's home was all but completely finished. Only some work on the outdoor kitchen and open air dining room and we'd be done. After service, I slipped into my bathing suit, complete with the "In-n-Out Burgers" logo emblazoned on the side. The pastor drove my motorbike back and forth, picking up as many people as he could to go to the watering hole.
3 of the former street boys and some young teenage girls would be getting baptized. A couple of young mothers in the church were joining them. The entire church had come out to the watering hole to see. Many villagers gathered around to watch, hear the singing, and the scripture readings.
I walked cautiously into the muddy watering hole, hoping not to find a snake or rogue stick. After about 20 feet out, I had reached a depth where I could get people fully immersed. Problem was, the mud was very soft and immediately I began to sink. It was on with the program, anyway. First the boys and girls came out, then the mothers. Then a few other church members came out. Then some people who just came over to watch came out to be baptized. I started

feeling like I was filming the baptism scene in "O Brother Where Art Though"!

The last to go in was my own daughter, Miriam. After calming her down about the state of the water, she was baptized joyously. As we exited, I had to get help freeing myself from the deep mud, into which I was up to my knees. Immediately, I felt itchy. Coming out of the water, my legs were covere with leeches. I looked like I was wearing knee-high socks made out of leeches. Immediately, without my request, kids were yanking them off of me. This only made the blood freely flow from where they were. So, I was whisked home looking like my legs had been shaved with a hacksaw.

There was really no time to slow down, though, because it was time to slaughter the goat we had purchased for this occasion and start cooking the feast to celebrate the new children's home.

People admired the new building, built the popular Kikuyu style with wooden posts, but cement in place of mud. They walked around chattering happily as they gnawed on the bones of the goat the kids were playing with just hours ago. Exactly one week later the kids were moving in onto their new beds, showing off their new belongings, sporting their new school uniforms. The excitement that was there the week before when we held the construction celebration was now doubled.

We walked the children to the church, their new mom and dad leading the way. An adventure in ministry and orphan care was on their horizon and they were eager to greet it. At the church, many people gathered to pray for the new family that was born that day. I preached on the challenges and blessings of caring for the fatherless. The church pastor, their new dad, spoke about accepting their challenges and striving for pure religion. We then presented the new Naeuwa Family to the congregation. They clapped and hollered in joy. Then the room was filled with prayer for the home, the kids, the parents, the new Matron (children's care provider and Auntie) and the future. After the prayers, everyone was back out in front of the new children's home feasting on roasted meat with chapati and greens.

The final days we were there were spent coaching the new ministry director through any challenges and concerns he could think up. We were also finishing work on the outdoor kitchen, that would be finished the day before our departure on October 23rd, 2014.

Our final Sunday was spent tag-team preaching with the church minister and children's home director. The new parents for our little African Princess, Blessed, had come all the way to worship with us and take custody of the girl God had entrusted us with for a short time. We said goodbye to them and to our faithful helper in the home and ministry, Philis, the next day.

Slowly our packing was coming together. Projects were being finished. People were leaving. Then on an early Wednesday morning, we loaded the last of our things into a car and started the long, bumpy journey toward home. Waving with tears in our eyes, we said goodbye to place we called home for the last 6 months. We said goodbye to the children we'd seen on the streets, in our home, and now in their new family. Then we drove into the morning mist on the savanna. There was nothing else to see, as we drove one last time, among the herds of zebras in the mist.

Nate receives a flower of welcome from tribal elder.

The Bantas' tented home on the savanna.

Posts for the new children's home are delivered.

Putting up the trusses.

Pouring the kitchen floor.

Micah helping plaster the walls.

The completed home.

Baptisms at the watering hole.

From Top Left:

1- Stacy with Robert & Philip.

2- Nate & Blessed.

3- Daniel with Robert & Sheila.

4- Miriam & Sheila.

5- Micah & the boys.

6

Meet the Original Family

"and I will be a father to you,
  and you shall be sons and daughters to me,
says the Lord Almighty."
- 2 Corinthians 6:18 (ESV)

Robert was the first street boy we helped. The director of our partner NGO, Christ Domain Ministries, and the director of the soon-to-be Naeuwa Family, were on scene when he was picked up from the storefront, suffering from hypothermia. His mother was an alcoholic and mentioned other addictions. She was the first to kick Robert out of their home. His eating and clothing needs were too expensive. When he wandered back home, his father beat him badly and threw him back out on the streets, telling him never to return.

Our first goal with Robert was to try and reunite him with his family. This was seen as impossible from the start. We found his mother in jail, awaiting sentencing for something she had done. She had no desire to try and reunite, but was quick to sign custody over to the Naeuwa Family. The father, on the other hand, was much worse. His father's mother tried to intervene and say it was better for David, the Naeuwa Family Director, and me to not have come and that we should just leave. We insisted we at least wanted the father to sign over custody. His father stormed out of his workplace and declared that the boy was not his son, just a worthless street boy and that he'd harm us if we even tried to talk to him about that boy again.

Robert was difficult, at first. Running away, having difficulty becoming "just a boy" after many months on the street. Today the 9 year-old is a happy, active member of the Naeuwa Family.

Manuel and Shiela are sibling, former AIDS orphans. A few years before our arrival, their father had died of AIDS. Their mother came looking for help from the director of Domain Christ Ministries. She said she would help them, if in turn they helped her with her then-baby girl and young boy. A few days after the arrangement was made, she disappeared. Today, she is seen every now and then serving as a prostitute in town.

The director of Domain Christ Ministries first voiced interest in starting a children's home because of these two children and others he'd met like them. He was unable to care for them due to he and his wife's age and living arrangements. Sheila was infested with parasites that took months to resolve and heal completely. Now she has embraced being the little girl in the family. Her brother, Manuel, immediately took to being in the new family; as it meant school, regular meals, being with his sister, and having his own bed.

Philip was found by members of one of the partner churches there in Samburu County. He was known to them because he would occasionally attend church in their village. His father died from violence between tribes. His mother was quick to run off with another man and was never seen again. He was left in the care of his teenaged sister and grandmother, who herself was over 80, blind, and living in poverty. He managed well until his sister became ill with malaria. Soon, malaria had taken his last living relative who was capable of caring for him. The church did what they could for the grandmother, but Philip was still malnourished and took to the streets during the day to beg for food. He became feared among the other street boys, but would still often fight to prove his worth.

On the day he became part of the Naeuwa family, two church members found him covered in dirt, fighting with other street boys. He had tried finding food again that day, but was unsuccessful. He was only too eager to try having a family of his own.

Philip proved to be the greatest challenge among the 3 former street boys. He was most inclined to fits of rage and deep sorrow. He was also the boy who would exhibit suicidal and destructive behaviors whenever I left the home on business.

Today, the young boy has adjusted very well. He has given his life to Christ and trusted in Him for internal healing. He has become a productive and caring member of the Naeuwa family, though his struggles with his past still haunt him.

Philip still goes and visits his grandmother as often as he can.

Rafael was the last boy to join us from the streets. He was abandoned there by his parents. He used his street smarts and cunning to keep himself going. He was among 5 street boys that I was working with to rehabilitate and move into the home. He showed so much progress that he was able to join us, just a few weeks before we turned the home over to the new dad and Director.

Rafael still continues to impress with his mature attitude and cunning. He speaks Swahili, Samburu, Turkana, and English. Though he is starting to forget his native Turkana, as there are few to speak it with in Samburu.

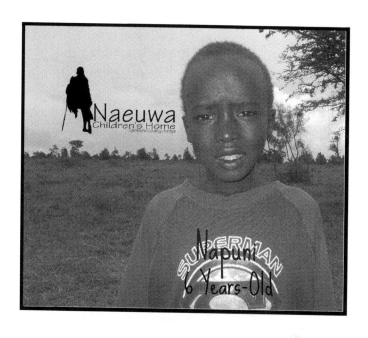

Six year-old Napuni came to the Naeuwa Family last. He was living with an ailing elderly couple who worried for his future, since they believe their own future to be so limited. They were told that if the former street boys were rehabilitated enough for Napani to be their little brother, that he'd be welcome into the Naeuwa Family. The week before my family left Kenya, Napani was notified that he could join the family and to pack up all his clothes and come move in. The couple came with him and a little sack with one shirt. Napuni lost both his parents to AIDS and was taken in by the kind, Christian couple. But they never had the money to feed him more than once a day or to take him to school. Napuni had spent his early childhood helping care for the elderly couple. He never left to play with friends and the Naeuwa Family and their neighbors were the first children he'd ever been able to play with.

Napuni never smiled the whole week we had him. His new brothers tried in vain to teach him for the sake of the picture above.

When I returned to check in on the home six months later, I pulled into the yard and found Napuni sitting on the seesaw, smiling from ear to ear as he was giggling away with some of the neighbor kids and Sheila.

From Top Left:

1- Robert & Manuel

2- Sheila celebrates her new bed & belongings.

3- Philip goofing off

4- The foster kids, Micah, & Nate on the "mini van".

___Afterwards___

Going Home and Continuing On

And I am sure of this, that he who began a good work in you will bring it to completion at the day of Jesus Christ. - Philippians 1:6

Since returning from Kenya, the Naeuwa Family has continued to thrive. Fairdale Christian Church in Fairdale, Kentucky has adopted them as their church's children's home and faithfully support the ministry of the Naeuwa Family. All the children have also found personal sponsors through the kindness of the members at Fairdale Christian Church.

The Naeuwa Family has also grown. Three new former street children have begun rehabilitation and have joined the home. They came into the family at the very young ages of five, five, and seven.

My family has moved on to helping start new families in India and Southeast Asia. We have partnered with the Faith Family in Southeast Asia and the Life Care Family in northeast India. As of writing this, the Ellen Family is nearly ready to open their doors and arms to the neediest children in Odisha, India.

___All About___
Author and Organization

The Author

I am a fourth generation missionary. My great-grandparents, Norton and Lois Bare, were missionaries to China and Tibet in the 1920s through 1930s. My grandparents, Archie and Marguerite Fairbrother were missionaries to India and Hong Kong. My parents, Larry and Ellen Banta, were missionaries to Kenya and Mexico. While I just go anywhere God sends me to help start up a family for orphans and street kids. I studied at Boise Bible College. Before missions re-entered my life, I was a children's minister at Eagle Christian Church in Eagle, ID, Caldwell Christian Church in Caldwell, ID, and Shively Christian Church in Louisville, KY. Along with my faithful and beautiful wife, Stacy, we have been serving once again with Child Help International since April 2012. We have 4 awesome kids who bring joy to our lives and help us out some, too. Aside from leading CHI (TEN18), I am the pastor of Sunrise Christian Church in Ontario, Oregon.

The Organization

I created Child Help International (Now the TEN18 Foundation) in 2001 as a vehicle to evangelize underprivileged children. The focus was on short-term mission trips to put on evangelistic programs for children near Tijuana, Mexico. Apart from that program, CHI started another spectrum, orphan care evangelism (taking care of orphans in a home that emphasizes faith in Christ, love for the church, and service.) In 2002, I was able to oversee the start of a children's home near Kitale, Kenya.

In 2003, I resigned from CHI to enter church ministry full-time and the organization survived as solely a board-run organization.

In 2012, "The Rebirth of Child Help International" began when Stacy and I resigned our positions in Kentucky and renewed our positions with CHI. The board undertook a complete revisioning of the organization to focus on a reproducible orphan care model that my father, Larry Banta, and I conceived: To assist the local church to initiate an orphan care project that is modeled after a Christian family. CHI is the empowering force to begin the work and sustain it in a way that helps the children succeed eternally, the US partners are the catalyst that drives the ministry, and God is the provider of the wisdom and guidance to direct the ministry advancement.

You can learn more about the TEN18 Foundation at our website:

www.childhelpinternational.com

If you are ever in the Caldwell, Idaho region, be sure to stop by CHI Coffee, our coffee house that raises capital to start more families like the Naeuwa Family! Learn more at www.chicoffee.org.

Made in the USA
Columbia, SC
28 September 2022